YOUR KNOWLEDGE HAS VALUE

AF 136904

- We will publish your bachelor's and
  master's thesis, essays and papers

- Your own eBook and book -
  sold worldwide in all relevant shops

- Earn money with each sale

Upload your text at www.GRIN.com
and publish for free

An Analytical Technique to Solve the System of Non-Linear Korteweg–De Vries Equation (KdV), FPDE

Danish Ali Raza

Bibliographic information published by the German National Library:

The German National Library lists this publication in the National Bibliography; detailed bibliographic data are available on the Internet at http://dnb.dnb.de.

ISBN: 9783346490582
This book is also available as an ebook.

© GRIN Publishing GmbH
Nymphenburger Straße 86
80636 München

Print and binding: Books on Demand GmbH, Norderstedt, Germany
Printed on acid-free paper from responsible sources.

The present work has been carefully prepared. Nevertheless, authors and publishers do not incur liability for the correctness of information, notes, links and advice as well as any printing errors.

GRIN web shop: https://www.grin.com/document/1127621

An Analytical Technique to solve the system of non-linear KdV, FPDE

(By Laplace transform — new novel analytic method)

By

Ijlal Hussain and Danish Ali Raza

Department of Mathematical Science

Federal Urdu University of Arts, Sciences & Technology

Dedicated to our beloved teacher

Dr. Hafiz Syed Hussain

ABSTRACT

Fractional Calculus (FC) goes back to the dawn of the theory of differential calculus. however, the application of FC just came up in the last two decades, due to the breakthrough in the area of chaos that disclose subtle connections with the FC concepts.

In this analysis, the explicit solutions to a generalized Korteweg–de Vries equation (KdV for short) with initial condition are calculated by using the new Analytical method.

Korteweg–de Vries (KdV) equation, which are characterized by the solitary wave solutions of the classical nonlinear equations that lead to solitons. Here, the classical nonlinear equations of interest usually admit for the existence of a special type of the traveling wave solutions which are either solitary waves or solitons.

Conseder the general Kortweg − de Vries(KdV)

$$D^\gamma u(x_1, t_1) + Lu(x_1, t_1) + Nu(x_1, t_1) = q(x_1, t_1), \ x_1, t_1 \geq 0, m - 1 < \gamma < m.$$

The initial condition is

$$u(x_1, o) = k_0(x_1), 0 < \gamma \leq 1, t_1 > 0.$$

Keywords: *Fractional calculus; fractional integral; Fractional differential equations Korteweg–de Vries equation; caputo operator.*

ACKNOWLEDGMENTS

In the name of Allah, the most Merciful, most Sympathetic.

We thank ALLAH for granting us with health, patience, and proficiency to complete this thesis and without ALLAH's elegance, we couldn't have done it. So, to ALLAH returns all the honor and gratitude.

We would like to thank our supervisor Dr. Marium Sultana for her dedicated support and guidance. Her continuously provided encouragement and was always willing and enthusiastic to assist in any way she could throughout the research project.

During this project, we acquired many valuable skills, and we hope that in the years to come, those skills will be put to good use.

Last but not the least; We would like to thank our colleagues for supporting us physically and spiritually throughout our life. May Allah reward them with the best reward.

Contents

List of Symbols

List of Abbreviations

Abbreviations	*Denoted*
KdV	*Kortweg − de Vries*
R-L	*Riemann-Liouville*
FDEs	*Fractional Differential Equations*
FC	*Fractional Calculus*

Chapter 1:

Introduction

The idea of integration and differentiation is familiar to all who have studied elementary calculus. We know, for instance, that if $f(x) = x^2$ then integrating $f(x)$ to the 1st order results in $\int f(x)dx = \frac{1}{3}x^3 + c_1$ and integrating the same function to the 2nd order results $\int[\int f(x)dx]\,dx = \frac{1}{12}x^4 + xc_1 + c_2$. Similarly, $\frac{df(x)}{dx} = 2x$, and $\frac{d^2 f(x)}{dx^2} = 2$. However, what if we desire to integrate our function $f(x)$ to the $\left(\frac{1}{2}\right)^{th}$ order, or find it's $\left(\frac{1}{2}\right)^{th}$ order derivative? How could we define our operations? Better still, would our results have a meaning or an application comparable to that of the familiar integer order operations?

1.1 *Origin of the fractional derivative*

The theory of integer order derivative and integral are common. The derivative $\frac{d^n y}{dx^n}$ express the changes of variable y with respect to variable x, and has a profound physical background. The present difficulty is how to generalize n into a fraction, even a complex number.

This long-established problem can be dated back to the letter from L'Hopital to Leibniz in 1695, in which it is asked what the derivative $\frac{d^n y}{dx^n}$ is when $n = \frac{1}{2}$. In the same year, the derivative of general order was introduced in the letter from Leibniz to J. Bernoulli. The issue was also considered by Euler (1730), Lagrange (1849) et al, and gave some material insights. In 1812, by using the concept of integral, Laplace provided a definition of fractional derivate. When $y = x^m$, employing the gamma function.

$$\frac{d^n y}{dx^n} = \frac{\Gamma(m+1)}{\Gamma(m-n+1)} x^{m-n}, \quad m \geq n,$$

Was derived by Lacroix, which gives

$$\frac{d^{\frac{1}{2}} y}{dx^{\frac{1}{2}}} = \frac{2\sqrt{x}}{\sqrt{\pi}}.$$

When $y = x$ and $n = \frac{1}{2}$. This is consistent with the so-called Riemann-Liouville fractional derivative.

Soon later, Fourier (1822) gave the definition of fractional derivative through the so-called Fourier transform.

In 1930s, Liouville, possibly motivated by Fourier and Abel, made a series of work in the branch of fractional derivative, and successfully applied them into the potential theory. Since

1.2 Historical Background

Over one hundred and fifty years ago, while conducting experiments to determine the most efficient design for canal boats, a young Scottish engineer named John Scott Russell (1808-1882) made a remarkable scientific discovery.

It was not until the mid-1960's when applied scientists began to use modern digital computers to study nonlinear wave propagation that the soundness of Russell's early ideas began to be appreciated. He viewed the solitary wave as a self-sufficient dynamic entity, a "thing" displaying many properties of a particle. From the modern perspective it is used as a constructive element to formulate the complex dynamical behavior of wave systems throughout science.

From hydrodynamics to nonlinear optics, from plasmas to shock waves, from tornados to the Great Red Spot of Jupiter, from the elementary particles of matter to the elementary particles of thought. The phenomenon described by Russell can be expressed by a non-linear Partial Differential Equation of the third order.

A partial differential equation (PDE) is a mathematical equation which contains an unknown function of more than one variable as well as some derivatives of that function with respect to the different independent variables. In practical applications where the PDE describes a dynamic process one of the variables has the meaning of the time (hence denoted by t) and the other (normally only up to 3) variable have the meaning of the space (hence denoted by x, y and z) We consider the simplest case in only one space variable x. So, we are looking for a function u depending on the variables x and t,

i.e., u(x, t) which describes the elongation of the wave at the place x at time t.

Using the typical short denotations as

$$u_t(x,t) = \frac{\partial u(x,t)}{\partial t}; \ u_x(x,t) = \frac{\partial u(x,t)}{\partial x}; \ u_{xx}(x,t) = \frac{\partial^2 u(x,t)}{\partial x^2}; \ u_{xxx}(x,t) = \frac{\partial^3 u(x,t)}{\partial x^3}$$

the problem can be formulated as:

$$u_t(x,t) + 6u(x,t)u_x(x,t) + u_{xxx}(x,t) = 0$$

This is the Korteweg-de Vries Equation (KdV) which is nonlinear because of the product shown in the second summand and which is of third order because of the third derivative as highest in the third summand. The factor 6 is just a scaling factor to make solutions easier to describe.

1.3 *Applications*

Mathematical modeling of real-life problems generally results in fractional differential equations and many other problems require special functions of mathematical physics as well as their extensions and generalizations in one or more variables.

Additionally, most physical phenomena of fluid dynamics, quantum mechanics, electricity, ecological systems, and many other models are controlled within their domain of validity by fractional order PDEs. Therefore, it becomes increasingly important to be familiar with all traditional and recently advance methods for solving fractional order PDEs and the implementations of these methods.

Right now, the use of fractional order partial differential equation in real-physical systems is commonly encountered in the fields of science and engineering.

The efficient computational tools are required for analytical and numerical approximations of such physical models. The present issue has addressed recent trends and developments regarding the analytical and numerical methods that may be used in the fractional order dynamical systems.

Eventually, it may be expected that the present special issue would certainly helpful to explore the researchers with their new arising fractional order problems and elevate the efficiency and accuracy of the solution methods for those problems in use nowadays.

Chapter 2:
Definations and Preliminary Notion

2.1 The Gamma Function

In the integer-order calculus the factorial plays an important role because it is one of the most fundamental combinatorial tools. The Gamma function has the same importance in the fractional-order calculus and it is basically given by integral.

$$\Gamma(z) = \int_0^\infty e^{-t} t^{z-1} dt \qquad (2.1.1)$$

Lemma 2.1.1
Prove that $\Gamma(z + 1) = z\Gamma(z)$.

proof:

by equation (2.1.1)

$$\Gamma(z) = \int_0^\infty e^{-t} t^{z-1} dt$$

$$\Rightarrow \quad \Gamma(z + 1) = \int_0^\infty e^{-t} t^z dt = -e^{-z} t^z \Big|_{t=0}^{t=\infty} + z \int_0^\infty e^{-t} t^{z-1} dt = z\Gamma(z)$$

2.2 The Laplace Transform:

The Laplace transform of a function $h(t)$ defined for all real numbers $t \geq 0$, is the function $H(s)$, which is a unilateral transform defined by

$$H(s) = \int_0^\infty h(t) e^{-st} dt,$$

Property 2.2.1: *The laplace transform of* $h(t), t > 0$ *is defined by*

$$H(s) = \mathcal{L}[h(t)] = \int_0^\infty e^{-st} h(t) dt.$$

The Laplace transform of function $h(t) = t^\alpha$ is given for α as $non-integer$

order $n - 1 < \alpha \leq n$

$$\mathcal{L}[h(t)] = \frac{\Gamma(\alpha + 1)}{s^{\alpha+1}}$$

Property 2.2.2: *The Laplace transform in term of convolution is given by*

6

$$\mathcal{L}[h_1 * h_2] = \mathcal{L}[h_1] * \mathcal{L}[h_2]$$

$here, h_1 * h_2, define\ the\ convolution\ h_1\ and\ h_2,$

$$(h_1 * h_2)t = \int_0^\tau h_1(\tau)h_2(t - \tau)dt$$

2.3 The Riemann-Liouville

The Riemann–Liouville integral is defined by

$R - L\ fractional\ integral:$

$$I_x^\gamma g(x) = \begin{cases} g(x), & if\ \gamma = 0 \\ \dfrac{1}{\Gamma(\gamma)} \displaystyle\int_0^x (x - v)^{\gamma-1} g(v)dv, & if\ \gamma > 0 \end{cases}$$

$where\ \Gamma\ denote\ the\ gamma\ function\ defined\ by$

$$\Gamma(z) = \int_0^\infty e^{-x} x^{z-1} dx, \qquad z \in \mathbb{C}$$

$When\ z = m \in \mathbb{N}, then\ gamma\ function\ is\ related\ to\ the\ factorial\ function$

$$\Gamma(m) = (m - 1)!$$

$if\ m = n + 1, then$

$$\Gamma(n + 1) = (n)!$$

2.4 Caputo fractional derivative

The Caputo derivative with fractional order that was proposed by Italian Caputo in 1967. This form of fractional derivative is given as

$The\ caputo\ operator\ of\ order\ \gamma\ for\ fractional\ derivative\ is\ given\ by\ the$

$following\ mathematical\ expression\ for\ n \in\ \mathbb{N}, x > 0, g \in \mathbb{C}_t, t \geq 1:$

$$D^\gamma g(x) = \frac{\partial^\gamma g(x)}{\partial t^\gamma} = \begin{cases} I^{n-\gamma}\left[\dfrac{\partial^\gamma g(x)}{\partial t^\gamma}\right], & if\ n - 1 < \gamma \leq n, m \in \mathbb{N}, \\ \dfrac{\partial^\gamma g(x)}{\partial t^\gamma} \end{cases}$$

Linearity Property: *if* $n-1 < \gamma \le n$ *with* $n \in \mathbb{N}$, λ
$\in \mathbb{C}$, *and functions* $f(x), g(x)$ *then caputo*

fractional derivative is a linear operator, i.e.

$$D^\gamma[\lambda f(x) + g(x)] = \lambda D^\gamma f(x) + D^\gamma g(x)$$

Hence , we required the subsequent properties given in next lemma.

Lemma 2. 4. 1:

if $n-1 < \gamma \le n$ *with* $n \in \mathbb{N}$ *and* $g \in \mathbb{C}_t$ *with* $t \ge -1$, *then*

$$I^\gamma I^a g(x) = I^{\gamma+a} g(x), \quad a, \gamma \ge 0.$$

$$I^\gamma I^\lambda = \frac{\Gamma(\lambda+1)}{\Gamma(\gamma+\lambda+1)} x^{\gamma+\lambda}, \gamma > 0, \lambda > -1, x > 0.$$

$$I^\gamma D^\gamma g(x) = g(x) - \sum_{k=0}^{n-1} g(x)^k (0)^+ \frac{x^k}{k!}, \quad for \; x > 0, n-1 < \gamma \le n.$$

Fractional derivative in term of Laplace transform is

$$\mathcal{L}\left(D_t^\gamma h(t)\right) = s^\gamma H(s) - \sum_{k=0}^{n-1} s^{\gamma-1-k} h^k(0),$$

where $H(s)$ *is the Laplace transform of* $h(t)$.

2.5 *Ideal of Fractional Laplac – new novel analytic method*

Now we will discuss Laplace transform – new novel analytic method for the solution of

FPDEs:

Conseder the general Kortweg – de Vries(KdV)

$$D^\gamma u(x_1, t_1) + Lu(x_1, t_1) + Nu(x_1, t_1) = q(x_1, t_1), \quad x_1, t_1 \ge 0, m-1 < \gamma < m. \tag{2.5.1}$$

where $D^\gamma = \frac{\partial^\gamma}{\partial t_1^\gamma}$ *the captuo Operator* $\gamma, m \in \mathbb{N}$, *where L and N are nonlinear functions, q is the*

source function.

The initial condition is

8

$$u(x_1, o) = k_0(x_1), 0 < \nu \leq 1, t_1 > 0. \tag{2.5.2}$$

Applying the laplace transform to Equation (2.5.1), we have

$$\mathcal{L}[D^\gamma u(x_1, t_1)] + \mathcal{L}[Lu(x_1, t_1)] + \mathcal{L}[Nu(x_1, t_1)] = \mathcal{L}[q(x_1, t_1)] \tag{2.5.3}$$

and using the differentiation property of Laplace transform, we get

$$s^\gamma \mathcal{L}[u(x_1, t_1)] - s^{\gamma-1} u(x_1, 0) = \mathcal{L}[q(x_1, t_1)] - \mathcal{L}[Lu(x_1, t_1)] - \mathcal{L}[Nu(x_1, t_1)],$$

$$s^\gamma \mathcal{L}[u(x_1, t_1)] = s^{\gamma-1} u(x_1, 0) + \mathcal{L}[q(x_1, t_1)] - \mathcal{L}[Lu(x_1, t_1)] - \mathcal{L}[Nu(x_1, t_1)],$$

$$\mathcal{L}[u(x_1, t_1)] = \frac{k_0(x_1)}{s} + \frac{\mathcal{L}[q(x_1, t_1)]}{s^\gamma} - \frac{\mathcal{L}[Lu(x_1, t_1) - Nu(x_1, t_1)]}{s^\gamma}. \tag{2.5.4}$$

Applying the inverse Laplace transform, in Equation (4)

$$u(x_1, t_1) = k_0(x_1) + \mathcal{L}^{-1}\left[\frac{\mathcal{L}[q(x_1, t_1)]}{s^\gamma} - \frac{\mathcal{L}[Lu(x_1, t_1) - Nu(x_1, t_1)]}{s^\gamma} \right]$$

Now applying new novel analytic method

The Taylor series

$$u(x_1, t_1) = a_0 + a_1 t_1 + a_2 \frac{t_1}{2!} + a_3 \frac{t_1}{3!} + a_4 \frac{t_1}{4!} + \cdots$$

where

$$a_0 = k_0(x_1)$$

$$a_1 = \frac{\partial u(x_1, 0)}{\partial t_1}$$

$$a_2 = \frac{\partial^2 u(x_1, 0)}{\partial t_1^2}$$

$$a_3 = \frac{\partial^3 u(x_1, 0)}{\partial t_1^3}$$

.   .

.   .

.   .

.   .

$$a_n = \frac{\partial^n u(x_1, 0)}{\partial t_1^n}$$

so we get

$$u(x_1, t_1) = k_0(x_1) + \frac{\partial u(x_1, 0)}{\partial t_1} t_1 + \frac{\partial^2 u(x_1, 0)}{\partial t_1^2} \frac{t_1}{2!} + \frac{\partial^3 u(x_1, 0)}{\partial t_1^3} \frac{t_1}{3!} + \frac{\partial^4 u(x_1, 0)}{\partial t_1^4} \frac{t_1}{4!} + \cdots \frac{\partial^n u(x_1, 0)}{\partial t_1^n} \frac{t_1}{n!}$$

2.6 Generating an analytical approach:

In this section, we will discuss the basic ideas of constructing a novel analytical method.

Let us consider the initial value problem.

$$w_{tt}(x, t) = G(w_t, w, u_x, w_{xx}, \ldots), \qquad\qquad 2.6.1$$

With initial condition

$$w(x, 0) = h_0(x), \qquad\qquad w_t(x, 0) = h_1(x), \qquad\qquad 2.6.2$$

By using the integral for the two sides of equation (2.6.1) from 0 to t, we get

$$w_t(x, t) - w_t(x, 0) = \int_0^t F[w] dt.$$

$$w_t(x, t) - h_1(x) = \int_0^t F[w] dt.$$

Then

$$w_t(x, t) = h_1(x) + \int_0^t F[w] dt. \qquad\qquad 2.6.3$$

Where $G[w] = G(w_t, w, u_x, w_{xx}, \ldots)$.

Then, when the integral of two sides of equation (2.6.3) is used from 0 to t, we obtain

$$w(x, t) - w(x, 0) = h_1(x)t + \int \int_0^t G[w] dt\, dt.$$

$$w(x, t) - h_0(x) = h_1(x)t + \int \int_0^t G[w] dt\, dt.$$

Thus,

$$w(x, t) = h_0(x) + h_1(x)t + \int \int_0^t G[w] dt\, dt. \qquad\qquad 2.6.4$$

The Taylor series is extended for $G[w]$ about $t = 0$, which is

$$G[w] = G[w_0] + G'[w_0]t + G''[w_0]\frac{t^2}{2!} + G'''[w_0]\frac{t^3}{3!} + \cdots + G^{(n)}[w_0]\frac{t^n}{n!} \dots \qquad 2.6.5$$

Substituting equation (2.6.5) by equation (2.6.5), we get

$$(x,t) = h_0(x) + h_1(x)t + G[w_0]\frac{t^2}{2!} + G'[w_0]\frac{t^2}{2!} + G''[w_0]\frac{t^4}{4!} + \cdots + G^{(n-2)}[w_0]\frac{t^2}{2!} \cdots$$

$$= a_0 + a_1t + a_2\frac{t^2}{2!} + a_3\frac{t^2}{2!} + \cdots + a_n\frac{t^n}{n!} + \cdots \qquad\qquad 2.6.6$$

Where

$a_0 = h_0(x),$

$a_1 = h_1(x),$

$a_2 = G[w_0],$

$a_3 = G'[w_0]$

.

.

.

$a_n = G^{(n-2)}[w_0]$

Where n is the highest derivative of u. The formal of Equation (2.6.6) is expand Taylor's series for w about $t = 0$. This means

$a_0 = w(x,0),$

$a_1 = \dfrac{\partial}{\partial t}w(x,0),$

$a_2 = \dfrac{\partial^2}{\partial t^2}w(x,0),$

$a_3 = \dfrac{\partial^3}{\partial t^3}w(x,0),$

.

.

.

$n = \dfrac{\partial^n}{\partial t^n}w(x,0),$

2.7 Convergence analysis:

Consider the PDE in the following form

$$u(x,t) = G(u(x,t))$$

Where G is a nonlinear operator. The solution that obtained by the presented technique is equivalent to the following sequence

$$S_n = \sum_{i=0}^{n} w_i = \sum_{i=0}^{n} \delta_i \frac{(\Delta t)^i}{(i)!}$$

Theorem 2.7.1:

Let G be an operator from a Hilbert space H into H and w be the exact solution of equation (2.7). The approximate solution $S_n = \sum_{i=0}^{n} w_i = \sum_{i=0}^{n} \delta_i \frac{(\Delta t)^i}{(i)!}$ is converged to exact solution w, when is converged to exact solution w, when $\exists\, 0 \le \delta < 1, \|w_{i+1}\| \le \delta \|w_i\| \,\forall\, i \in \mathbb{N}\; \cup\{0\}$.

Proof:

We want to prove that $\{S_n\}_{n=0}^{\infty}$ is a converged Cauchy Sequence,

$$\|S_{n+1} - S_n\| = \|W_{n+1}\| \le \delta\|W_n\| \le \delta^2\|W_{n-1}\| \le ....\delta^n\|W_1\| \le \delta^{n+1}\|W_0\|$$

Now for $n, m \in \mathbb{N}, n \ge m$ we get

$$\begin{aligned}
\|S_n - S_m\| &= \|(S_n - S_{n-1}) + (S_{n-1} - S_{n-2}) + \cdots + (S_{m+1} - S_m)\| \\
&\le \|(S_n - S_{n-1})\| + \|(S_{n-1} - S_{n-2})\| + \cdots + \|(S_{m+1} - S_m)\| \\
&\le \delta^n\|W_0\| + \delta^{n-1}\|W_0\| + \cdots + \delta^{m+1}\|W_0\| \le (\delta^{m+1} + \delta^{m+2} + \cdots + \delta^n)\|W_0\| \\
&= \delta^{m+1}\frac{1 - \delta^{n-m}}{1 - \delta}\|W_0\|
\end{aligned}$$

Hence $\lim_{n,m\to\infty}\|S_n - S_m\| =$
0 *i.e.,* $\{S_n\}_{n=0}^{\infty}$ *is a Cauchy Sequence in Hilbert Space H. Thus there exist $S \in H$ such that* $\lim_{n\to\infty} S_n = S$ *where $W = S$*

Definition 2.7.2:

For every $n \in \mathbb{N} \cup \{0\}$, we define

$$\delta_n = \begin{cases} \dfrac{\|W_{n+1}\|}{\|W_n\|}, & \|W_n\| \ne 0 \\ 0, & otherwise \end{cases}$$

Corollary 2.7.3: *From theorem 2.7.1,*

$$\sum_{i=0}^{\infty} W_i = \sum_{i=0}^{\square} \delta_i \frac{(\Delta t)^i}{(i)!},$$

Is converged to exact solution w *when* $0 \le \delta_i < 1, i = 0,1,2, \dots$

Therefore, the convergence of analytical solutions is valid. Finally, the theoretical proofs for the analysis of convergence coincide with the computation results presented in the below figures and tables.

2.8 *Some test problems:*

Example 1. *Consider the nonlinear KdV system of time − fractional order.*

$$\frac{\partial^{\gamma} u}{\partial t_1^{\gamma}} = -a \frac{\partial^3 u}{\partial x_1^3} - 6au \frac{\partial u}{\partial x_1} + 6v \frac{\partial v}{\partial x_1} \,, \qquad (2.8.1)$$

$$\frac{\partial^{\gamma} u}{\partial t_1^{\gamma}} = -a \frac{\partial^3 v}{\partial x_1^3} - 3au \frac{\partial v}{\partial x_1} \,, \qquad 0 < \gamma < 1 \qquad (2.8.2)$$

With initial conditions

$$u(x_1, 0) = \eta^2 \, Sech^2 \left(\frac{c}{2} + \frac{\eta x_1}{2} \right), \quad v(x_1, 0) = \sqrt{\frac{a}{2}} \eta^2 \, Sech^2 \left(\frac{\alpha}{2} + \frac{\eta x_1}{2} \right)$$

Where the constat a is a wave veocity and η, α *are arbitrary constants.*

Taking Laplace transform of Equations (1.1) and (1.2)

$$\mathcal{L}\left[\frac{\partial^{\gamma} u}{\partial t_1^{\gamma}} \right] = \mathcal{L}\left[-a \frac{\partial^3 u}{\partial x_1^3} - 6au \frac{\partial u}{\partial x_1} + 6v \frac{\partial v}{\partial x_1} \right]$$

$$\mathcal{L}\left[\frac{\partial^{\gamma} u}{\partial t_1^{\gamma}} \right] = \mathcal{L}\left[-a \frac{\partial^3 v}{\partial x_1^3} - 3au \frac{\partial v}{\partial x_1} \right]$$

$$s^{\gamma} \mathcal{L}[u(x_1, t_1)] - s^{\gamma-1} u(x_1, 0)] = \mathcal{L}\left[-a \frac{\partial^3 u}{\partial x_1^3} - 6au \frac{\partial u}{\partial x_1} + 6v \frac{\partial v}{\partial x_1} \right]$$

$$s^{\gamma} \mathcal{L}[v(x_1, t_1)] - s^{\gamma-1} \mathbb{P}(x_1, 0)] = \mathcal{L}\left[-a \frac{\partial^3 v}{\partial x_1^3} - 3au \frac{\partial v}{\partial x_1} \right]$$

Applying inverse Laplace transform

$$u(x_1, t_1) = \mathcal{L}^{-1}\left[\frac{u(x_1,0)}{s} + \frac{1}{s^\gamma}\mathcal{L}\left[-a\frac{\partial^3 u}{\partial x_1^3} - 6au\frac{\partial u}{\partial x_1} + 6v\frac{\partial v}{\partial x_1}\right]\right]$$

$$v(x_1, t_1) = \mathcal{L}^{-1}\left[\frac{v(x_1,0)}{s} + \frac{1}{s^\gamma}\mathcal{L}\left[-a\frac{\partial^3 v}{\partial x_1^3} - 3au\frac{\partial v}{\partial x_1}\right]\right]$$

$$u(x_1, t_1) = \eta^2\,Sech^2\left(\frac{\alpha}{2} + \frac{\eta x_1}{2}\right) + \mathcal{L}^{-1}\left[\frac{1}{s^\gamma}\mathcal{L}\left[-a\frac{\partial^3 u}{\partial x_1^3} - 6au\frac{\partial u}{\partial x_1} + 6v\frac{\partial v}{\partial x_1}\right]\right]$$

$$v(x_1, t_1) = \sqrt{\frac{a}{2}}\eta^2\,Sech^2\left(\frac{\alpha}{2} + \frac{\eta x_1}{2}\right) + \mathcal{L}^{-1}\left[\frac{1}{s^\gamma}\mathcal{L}\left[-a\frac{\partial^3 v}{\partial x_1^3} - 3au\frac{\partial v}{\partial x_1}\right]\right]$$

Applying new novel analytic method

For a_1 of $u(x_1, t_1)$

$$a_1 = \frac{\partial}{\partial t_1}\left(u(x_1,t_1)\right) = \frac{\partial}{\partial t_1}\left[u(x_1,0) + \mathcal{L}^{-1}[\frac{1}{s^\gamma}\mathcal{L}\left[-a\frac{\partial^3 u_0}{\partial x_1^3} - 6au\frac{\partial u_0}{\partial x_1} + 6v\frac{\partial v_0}{\partial x_1}\right]\right]$$

$$a_1 = \frac{\partial}{\partial t_1}[u(x_1,0)] + \frac{\partial}{\partial t_1}\left[+\mathcal{L}^{-1}[\frac{1}{s^\gamma}\mathcal{L}\left[-a\frac{\partial^3 u_0}{\partial x_1^3} - 6au\frac{\partial u_0}{\partial x_1} + 6v\frac{\partial v_0}{\partial x_1}\right]\right]$$

$$a_1 = \frac{\partial}{\partial t_1}\left[\eta^2\,Sech^2\left(\frac{\alpha}{2} + \frac{\eta x_1}{2}\right)\right] + \frac{\partial}{\partial t_1}\left[\eta^5 a\,Tanh\left(\frac{\alpha}{2} + \frac{\eta x_1}{2}\right)Sech^2\left(\frac{\alpha}{2} + \frac{\eta x_1}{2}\right)\frac{t_1^\gamma}{\Gamma(\gamma+1)}\right]$$

$$a_1 = 0 + \left[\eta^5 a\,Tanh\left(\frac{\alpha}{2} + \frac{\eta x_1}{2}\right)Sech^2\left(\frac{\alpha}{2} + \frac{\eta x_1}{2}\right)\frac{\gamma t_1^{\gamma-1}}{\Gamma(\gamma+1)}\right]$$

$$a_1 = \eta^5 a\,Tanh\left(\frac{\alpha}{2} + \frac{\eta x_1}{2}\right)Sech^2\left(\frac{\alpha}{2} + \frac{\eta x_1}{2}\right)\frac{\gamma t_1^{\gamma-1}}{\Gamma(\gamma+1)}$$

Similarly For a_2 of $u(x_1, t_1)$

$$a_2 = \eta^5 a\,Tanh\left(\frac{\alpha}{2} + \frac{\eta x_1}{2}\right)Sech^2\left(\frac{\alpha}{2} + \frac{\eta x_1}{2}\right)\frac{\gamma(\gamma-1)t_1^{\gamma-2}}{\Gamma(\gamma+1)}$$

Similarly For a_3 of $u(x_1, t_1)$

$$a_3 = \eta^5 a\,Tanh\left(\frac{\alpha}{2} + \frac{\eta x_1}{2}\right)Sech^2\left(\frac{\alpha}{2} + \frac{\eta x_1}{2}\right)\frac{\gamma(\gamma-1)(\gamma-2)t_1^{\gamma-3}}{\Gamma(\gamma+1)}$$

$\vdots \qquad\qquad \vdots$

$\vdots \qquad\qquad \vdots$

$\vdots \qquad\qquad \vdots$

:⠀⠀⠀⠀⠀⠀:

Putting the value of a_1, a_2, a_3 ... ... _ ...⠀⠀⠀in taylor series

$$u(x_1, t_1) = \eta^2 \, Sech^2\left(\frac{\alpha}{2} + \frac{\eta x_1}{2}\right) + \left[\eta^5 a \, Tanh\left(\frac{\alpha}{2} + \frac{\eta x_1}{2}\right) Sech^2\left(\frac{\alpha}{2} + \frac{\eta x_1}{2}\right)\frac{\gamma t_1^{\gamma-1}}{\Gamma(\gamma+1)}\right] t_1$$

$$+ \left[\eta^5 a \, Tanh\left(\frac{\alpha}{2} + \frac{\eta x_1}{2}\right) Sech^2\left(\frac{\alpha}{2} + \frac{\eta x_1}{2}\right)\frac{\gamma(\gamma-1)t_1^{\gamma-2}}{\Gamma(\gamma+1)}\right]\frac{t_1^2}{2!}$$

$$+ \left[\eta^5 a \, Tanh\left(\frac{\alpha}{2} + \frac{\eta x_1}{2}\right) Sech^2\left(\frac{\alpha}{2} + \frac{\eta x_1}{2}\right)\frac{\gamma(\gamma-1)(\gamma-2)t_1^{\gamma-3}}{\Gamma(\gamma+1)}\right]\frac{t_1^3}{3!} + \ldots$$

$$u(x_1, t_1) = \eta^2 \, Sech^2\left(\frac{\alpha}{2} + \frac{\eta x_1}{2}\right)\left[1 + a r_\cdot{}^3 Tanh\left(\frac{\alpha}{2} + \frac{\eta x_1}{2}\right)\frac{t_1^\gamma}{\Gamma(\gamma+1)}\left(1 + \sum_{n=1}^{\infty}\frac{(\gamma-n)!}{(n+1)!}\right)\right]$$

$$u(x_1, t_1) = \eta^2 \, Sech^2\left(\frac{\alpha}{2} + \frac{\eta x_1}{2}\right)\left[1 + a r_i^\varepsilon Tanh\left(\frac{\alpha}{2} + \frac{\eta x_1}{2}\right)\frac{t_1^\gamma}{\Gamma(\gamma+1)}\right] \tag{2.8.3}$$

Now finding b_1 of $v(x_1, t_1)$

$$b_1 = \frac{\partial}{\partial t_1}\left(v(x_1, t_1)\right) = \frac{\partial}{\partial t_1}\left[v(x_1, 0) + \mathcal{L}^{-1}[\frac{1}{s^\gamma}\mathcal{L}\left[-a\frac{\partial^3 v_0}{\partial x_1^3} - 3au\frac{\partial v_0}{\partial x_1}\right]\right]$$

$$b_1 = \frac{\partial}{\partial t_1}[v(x_1, 0)] + \frac{\partial}{\partial t_1}\left[\mathcal{L}^{-1}[\frac{1}{s^\gamma}\mathcal{L}\left[-c\frac{\partial^3 v_0}{\partial x_1^3} - 3au\frac{\partial v_0}{\partial x_1}\right]\right]$$

$$b_1 = \frac{\partial}{\partial t_1}\left[\sqrt{\frac{a}{2}}\eta^2 \, Sech^2\left(\frac{\alpha}{2} + \frac{\eta x_1}{2}\right)\right] + \frac{\hat{a}}{\partial t_1}\left[\frac{\eta^5 a^{\frac{3}{2}}}{\sqrt{2}} \, Tanh\left(\frac{\alpha}{2} + \frac{\eta x_1}{2}\right) Sech^2\left(\frac{\alpha}{2} + \frac{\eta x_1}{2}\right)\frac{t_1^\gamma}{\Gamma(\gamma+1)}\right]$$

$$b_1 = 0 + \frac{\eta^5 a^{\frac{3}{2}}}{\sqrt{2}} \, Tanh\left(\frac{\alpha}{2} + \frac{\eta x_1}{2}\right) Sech^2\left(\frac{\alpha}{2} + \frac{\eta x_1}{2}\right)\frac{\gamma t_1^{\gamma-1}}{\Gamma(\gamma+1)}$$

$$b_1 = \frac{\eta^5 a^{\frac{3}{2}}}{\sqrt{2}} \, Tanh\left(\frac{\alpha}{2} + \frac{\eta x_1}{2}\right) Sech^2\left(\frac{\alpha}{2} + \frac{r_i x_1}{2}\right)\frac{\gamma t_1^{\gamma-1}}{\Gamma(\gamma+1)}$$

Similarly For b_2 of $v(x_1, t_1)$

$$b_2 = \frac{\eta^5 a^{\frac{3}{2}}}{\sqrt{2}} \, Tanh\left(\frac{\alpha}{2} + \frac{\eta x_1}{2}\right) Sech^2\left(\frac{\alpha}{2} + \frac{\eta x_1}{2}\right)\frac{\gamma(\gamma-1)t_1^{\gamma-2}}{\Gamma(\gamma+1)}$$

Similarly For b_3 of $v(x_1, t_1)$

$$b_3 = \frac{\eta^5 a^{\frac{3}{2}}}{\sqrt{2}} \, Tanh\left(\frac{\alpha}{2} + \frac{\eta x_1}{2}\right) Sech^2\left(\frac{\alpha}{2} + \frac{\eta x_1}{2}\right) \frac{\gamma(\gamma - 1)(\gamma - 2)t_1^{\gamma - 3}}{\Gamma(\gamma + 1)}$$

$$\vdots \qquad \qquad \vdots$$

$$\vdots \qquad \qquad \vdots$$

Putting the value of $b_1, b_2, b_3 \dots \dots \dots \dots$ in taylor serie

$$v(x_1, t_1) = \sqrt{\frac{a}{2}}\eta^2 \, Sech^2\left(\frac{\alpha}{2} + \frac{\eta x_1}{2}\right) + \left[\frac{\eta^5 a^{\frac{3}{2}}}{\sqrt{2}} \, Tanh\left(\frac{\alpha}{2} + \frac{\eta x_1}{2}\right) Sech^2\left(\frac{\alpha}{2} + \frac{\eta x_1}{2}\right) \frac{\gamma t_1^{\gamma - 1}}{\Gamma(\gamma + 1)}\right] t_1$$

$$+ \left[\frac{\eta^5 a^{\frac{3}{2}}}{\sqrt{2}} \, Tanh\left(\frac{\alpha}{2} + \frac{\eta x_1}{2}\right) Sech^2\left(\frac{\alpha}{2} + \frac{\eta x_1}{2}\right) \frac{\gamma(\gamma - 1)t_1^{\gamma - 2}}{\Gamma(\gamma + 1)}\right] \frac{t_1^2}{2!}$$

$$+ \left[\frac{\eta^5 a^{\frac{3}{2}}}{\sqrt{2}} \, Tanh\left(\frac{\alpha}{2} + \frac{\eta x_1}{2}\right) Sech^2\left(\frac{\alpha}{2} + \frac{\eta x_1}{2}\right) \frac{\gamma(\gamma - 1)(\gamma - 2)t_1^{\gamma - 3}}{\Gamma(\gamma + 1)}\right] \frac{t_1^3}{3!} + \cdots$$

$$v(x_1, t_1) = \sqrt{\frac{a}{2}}\eta^2 \, Sech^2\left(\frac{\alpha}{2} + \frac{\eta x_1}{2}\right)\left[1 + a\eta^3 Tanh\left(\frac{\alpha}{2} + \frac{\eta x_1}{2}\right) \frac{\gamma t_1^\gamma}{\Gamma(\gamma + 1)}\left(1 + \sum_{n=1}^{\infty} \frac{(\gamma - n)!}{(n + 1)!}\right)\right]$$

$$v(x_1, t_1) = \sqrt{\frac{a}{2}}\eta^2 \, Sech^2\left(\frac{\alpha}{2} + \frac{\eta x_1}{2}\right)\left[1 + a\eta^3 Tanh\left(\frac{\alpha}{2} + \frac{\eta x_1}{2}\right) \frac{\gamma t_1^\gamma}{\Gamma(\gamma + 1)}\right] \qquad (2.8.4)$$

Now fixed values $a = \eta = 0.5, \alpha = 1$ and fixed order $\gamma = 1$ in (2.6.3), (2.6.4)
Table: *Solution of LNAM for different values of g when $h = 0.001$ and Absolute Error (AE) of Example 1.*

x_1	x_1	$\gamma = 0.55$		$\gamma = 1$		$Ex(\gamma = 1)$		AE	
		u_{LNAM}	v_{LNAM}	u_{LNAM}	v_{LNAM}	u_{Ex}	v_{Ex}	$u_{Ex} - u_{LNAM}$	$v_{Ex} - v_{LNAM}$
	0.1	0.0173252	0.008738	0.017556	0.008778	0.017661	0.002791	1.05×10^{-4}	5.8×10^{-3}
-10	0.3	0.0170452	0.008661	0.017343	0.008671	0.017659	0.002787	3.2×10^{-4}	5.82×10^{-3}
	0.5	0.0168449	0.008606	0.017130	0.008565	0.017657	0.002784	5.27×10^{-4}	5.78×10^{-3}
	0.1	0.198412	0.098801	0.197184	0.098589	0.196617	0.031096	5.67×10^{-4}	6.74×10^{-2}
0	0.3	0.199907	0.099212	0.198316	0.099157	0.196628	0.031114	1.68×10^{-3}	6.80×10^{-2}
	0.5	0.200975	0.099505	0.199451	0.099725	0.196640	0.031131	2.81×10^{-3}	6.859×10^{-2}
	0.1	0.002515	0.001246	0.002485	0.001240	0.002466	0.000390	1.9×10^{-5}	8.5×10^{-4}
10	0.3	0.002555	0.001257	0.002512	0.001257	0.002466	0.001256	4.6×10^{-5}	1×10^{-4}
	0.5	0.002584	0.122656	0.002543	0.001271	0.002467	0.000391	7.6×10^{-5}	8.8×10^{-4}

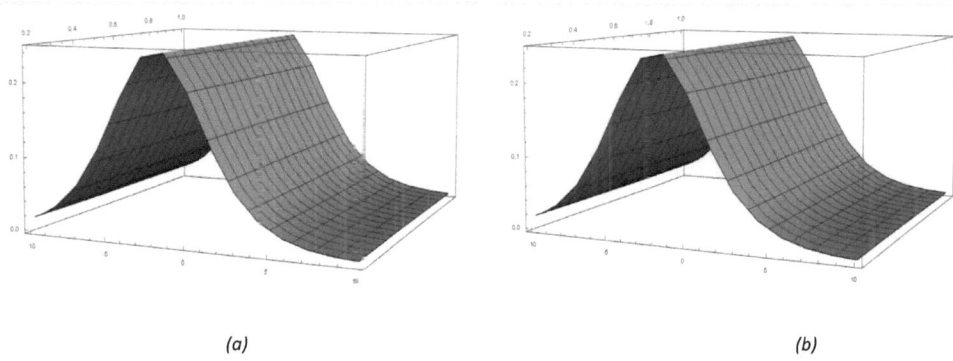

(a)   (b)

Solution of (a) $u(x_1, t_1)$ and (b) $v(x_1, t_1)$ of Example 1 at $\gamma = 1$

Example 2. *Consider the nonlinear dispersive long wave system of time −*
fractional order.

$$\frac{\partial^\gamma u}{\partial t_1^\gamma} = -\frac{\partial v}{\partial x_1} - \frac{1}{2}\frac{\partial u^2}{\partial x_1} , \qquad (2.8.5)$$

$$\frac{\partial^\gamma u}{\partial t_1^\gamma} = -\frac{\partial u}{\partial x_1} - \frac{\partial^3 v}{\partial x_1^3} - \frac{\partial uv}{\partial x_1} , \qquad 0 < \gamma < 1 \qquad (2.8.6)$$

With initial conditions

$$u(x_1, 0) = a\left[Tanh\left(\frac{\eta}{2} + \frac{ax_1}{2}\right) + 1\right], \quad v(x_1, 0) = -1 + \frac{1}{2}a^2 Sech^2\left(\frac{\eta}{2} + \frac{ax_1}{2}\right).$$

Taking Laplace transform of Equations (2.1) *and* (2.2)

$$\mathcal{L}\left[\frac{\partial^\gamma u}{\partial t_1^\gamma}\right] = \mathcal{L}\left[-\frac{\partial v}{\partial x_1} - \frac{1}{2}\frac{\partial u^2}{\partial x_1}\right]$$

17

$$\mathcal{L}\left[\frac{\partial^\gamma u}{\partial t_1^\gamma}\right] = \mathcal{L}\left[-\frac{\partial u}{\partial x_1} - \frac{\partial^3 v}{\partial x_1^3} - \frac{\partial uv}{\partial x_1}\right]$$

$$s^\gamma \mathcal{L}[u(x_1, t_1)] - s^{\gamma-1}[u(x_1, 0)] = \mathcal{L}\left[-\frac{\partial v}{\partial x_1} - \frac{1}{2}\frac{\partial u^2}{\partial x_1}\right]$$

$$s^\gamma \mathcal{L}[v(x_1, t_1)] - s^{\gamma-1}[v(x_1, 0)] = \mathcal{L}\left[-\frac{\partial u}{\partial x_1} - \frac{\partial^3 v}{\partial x_1^3} - \frac{\partial uv}{\partial x_1}\right]$$

Applying inverse Laplace transform

$$u(x_1, t_1) = \mathcal{L}^{-1}\left[\frac{u(x_1, 0)}{s} + \frac{1}{s^\gamma}\mathcal{L}\left[-\frac{\partial v}{\partial x_1} - \frac{1}{2}\frac{\partial u^2}{\partial x_1}\right]\right]$$

$$v(x_1, t_1) = \mathcal{L}^{-1}\left[\frac{v(x_1, 0)}{s} + \frac{1}{s^\gamma}\mathcal{L}\left[-\frac{\partial u}{\partial x_1} - \frac{\partial^3 v}{\partial x_1^3} - \frac{\partial uv}{\partial x_1}\right]\right]$$

$$u(x_1, t_1) = a\left[Tanh\left(\frac{\eta}{2} + \frac{ax_1}{2}\right) + 1\right] + \mathcal{L}^{-1}\left[\frac{1}{s^\gamma}\mathcal{L}\left[-\frac{\partial v}{\partial x_1} - \frac{1}{2}\frac{\partial u^2}{\partial x_1}\right]\right]$$

$$v(x_1, t_1) = -1 + \frac{1}{2}a^2 Sech^2\left(\frac{\eta}{2} + \frac{ax_1}{2}\right) + \mathcal{L}^{-1}\left[\frac{1}{s^\gamma}\mathcal{L}\left[-\frac{\partial u}{\partial x_1} - \frac{\partial^3 v}{\partial x_1^3} - \frac{\partial uv}{\partial x_1}\right]\right]$$

Applying new novel analytic method

For a_1 of $u(x_1, t_1)$

$$a_1 = \frac{\partial}{\partial t_1}\left(u(x_1, t_1)\right) = \frac{\partial}{\partial t_1}\left[u(x_1, 0) + \mathcal{L}^{-1}\left[\frac{1}{s^\gamma}\mathcal{L}\left[-\frac{\partial v_0}{\partial x_1} - \frac{1}{2}\frac{\partial u_0^2}{\partial x_1}\right]\right]\right]$$

$$a_1 = \frac{\partial}{\partial t_1}[u(x_1, 0)] + \frac{\partial}{\partial t_1}\left[+\mathcal{L}^{-1}\left[\frac{1}{s^\gamma}\mathcal{L}\left[-\frac{\partial v_0}{\partial x_1} - \frac{1}{2}\frac{\partial u_0^2}{\partial x_1}\right]\right]\right]$$

$$a_1 = \frac{\partial}{\partial t_1}\left[a\left[Tanh\left(\frac{\eta}{2} + \frac{ax_1}{2}\right) + 1\right]\right] + \frac{\partial}{\partial t_1}\left[-\frac{a^2}{2}Sech^2\left(\frac{\eta}{2} + \frac{ax_1}{2}\right)\frac{t_1^\gamma}{\Gamma(\gamma + 1)}\right]$$

$$a_1 = 0 + \left[-\frac{a^2}{2}Sech^2\left(\frac{\eta}{2} + \frac{ax_1}{2}\right)\frac{\gamma t_1^{\gamma-1}}{\Gamma(\gamma + 1)}\right]$$

$$a_1 = -\frac{a^2}{2}Sech^2\left(\frac{\eta}{2} + \frac{ax_1}{2}\right)\frac{\gamma t_1^{\gamma-1}}{\Gamma(\gamma + 1)}$$

Similarly For a_2 of $u(x_1, t_1)$

$$a_2 = -\frac{a^2}{2} Sech^2 \left(\frac{\eta}{2} + \frac{ax_1}{2}\right) \frac{\gamma(\gamma - 1)t_1^{\gamma-2}}{\Gamma(\gamma + 1)}$$

Similarly For a_3 of $u(x_1, t_1)$

$$a_3 = -\frac{a^2}{2} Sech^2 \left(\frac{\eta}{2} + \frac{ax_1}{2}\right) \frac{\gamma(\gamma - 1)(\gamma - 2)t_1^{\gamma-3}}{\Gamma(\gamma + 1)}$$

$\vdots \qquad\qquad \vdots$

$\vdots \qquad\qquad \vdots$

$\vdots \qquad\qquad \vdots$

$\vdots \qquad\qquad \vdots$

Putting the value of $a_1, a_2, a_3 \ldots \ldots \ldots \ldots$ in taylor series

$$u(x_1, t_1) = a \left[Tanh\left(\frac{\eta}{2} + \frac{ax_1}{2}\right) + 1\right] + \left[-\frac{a^2}{2} Sech^2 \left(\frac{\eta}{2} + \frac{ax_1}{2}\right) \frac{\gamma t_1^{\gamma-1}}{\Gamma(\gamma + 1)}\right] t_1$$

$$+ \left[-\frac{a^2}{2} Sech^2 \left(\frac{\eta}{2} + \frac{ax_1}{2}\right) \frac{\gamma(\gamma - 1)t_1^{\gamma-2}}{\Gamma(\gamma + 1)}\right] \frac{t_1^2}{2!}$$

$$+ \left[-\frac{a^2}{2} Sech^2 \left(\frac{\eta}{2} + \frac{ax_1}{2}\right) \frac{\gamma(\gamma - 1)(\gamma - 2)t_1^{\gamma-3}}{\Gamma(\gamma + 1)}\right] \frac{t_1^3}{3!} + \ldots$$

$$u(x_1, t_1) = a \left[Tanh\left(\frac{\eta}{2} + \frac{ax_1}{2}\right) + 1\right] - \frac{a^2}{2} Sech^2 \left(\frac{\eta}{2} + \frac{ax_1}{2}\right) \left(\frac{\gamma t_1^{\gamma}}{\Gamma(\gamma + 1)}\right) \left[1 + \sum_{n=1}^{\infty} \frac{(\gamma - n)!}{(n + 1)!}\right]$$

$$u(x_1, t_1) = a \left[Tanh\left(\frac{\eta}{2} + \frac{ax_1}{2}\right) + 1\right] - \frac{a^2}{2} Sech^2 \left(\frac{\eta}{2} + \frac{ax_1}{2}\right) \left(\frac{\gamma t_1^{\gamma}}{\Gamma(\gamma + 1)}\right) \qquad (2.8.7)$$

Now finding b_1 of $v(x_1, t_1)$

$$b_1 = \frac{\partial}{\partial t_1} \left(v(x_1, t_1)\right) = \frac{\partial}{\partial t_1} \left[v(x_1, 0) + \mathcal{L}^{-1}\left[\frac{1}{s^{\gamma}} \mathcal{L}\left[-\frac{\partial u_0}{\partial x_1} - \frac{\partial^3 v_0}{\partial x_1^3} - \frac{\partial u_0 v_0}{\partial x_1}\right]\right]\right]$$

$$b_1 = \frac{\partial}{\partial t_1} [v(x_1, 0)] + \frac{\partial}{\partial t_1} \left[\mathcal{L}^{-1}\left[\frac{1}{s^{\gamma}} \mathcal{L}\left[-\frac{\partial u_0}{\partial x_1} - \frac{\partial^3 v_0}{\partial x_1^3} - \frac{\partial u_0 v_0}{\partial x_1}\right]\right]\right]$$

$$b_1 = \frac{\partial}{\partial t_1}\left[-1 + \frac{1}{2}a^2 Sech^2\left(\frac{\eta}{2} + \frac{ax_1}{2}\right)\right] + \frac{\partial}{\partial t_1}\left[\frac{a^3}{2}Sinh\left(\frac{\eta}{2} + \frac{ax_1}{2}\right)Sech^3\left(\frac{\eta}{2} + \frac{ax_1}{2}\right)\frac{t_1^\gamma}{\Gamma(\gamma+1)}\right]$$

$$b_1 = 0 + \frac{a^3}{2}Sinh\left(\frac{\eta}{2} + \frac{ax_1}{2}\right)Sech^3\left(\frac{\eta}{2} + \frac{ax_1}{2}\right)\frac{\gamma t_1^{\gamma-1}}{\Gamma(\gamma+1)}$$

$$b_1 = \frac{a^3}{2}Sinh\left(\frac{\eta}{2} + \frac{ax_1}{2}\right)Sech^3\left(\frac{\eta}{2} + \frac{ax_1}{2}\right)\frac{\gamma t_1^{\gamma-1}}{\Gamma(\gamma+1)}$$

Similarly For b_2 of $v(x_1, t_1)$

$$b_2 = \frac{a^3}{2}Sinh\left(\frac{\eta}{2} + \frac{ax_1}{2}\right)Sech^3\left(\frac{\eta}{2} + \frac{ax_1}{2}\right)\frac{\gamma(\gamma-1)t_1^{\gamma-2}}{\Gamma(\gamma+1)}$$

Similarly For b_3 of $v(x_1, t_1)$

$$b_3 = \frac{a^3}{2}Sinh\left(\frac{\eta}{2} + \frac{ax_1}{2}\right)Sech^3\left(\frac{\eta}{2} + \frac{ax_1}{2}\right)\frac{\gamma(\gamma-1)(\gamma-2)t_1^{\gamma-3}}{\Gamma(\gamma+1)}$$

$$\vdots \qquad\qquad \vdots$$

$$\vdots \qquad\qquad \vdots$$

$$\vdots \qquad\qquad \vdots$$

Putting the value of b_1, b_2, b_3 ... ... in taylor serie

$$v(x_1, t_1) = -1 + \frac{1}{2}a^2 Sech^2\left(\frac{\eta}{2} + \frac{ax_1}{2}\right) + \left[\frac{a^3}{2}Sinh\left(\frac{\eta}{2} + \frac{ax_1}{2}\right)Sech^3\left(\frac{\eta}{2} + \frac{ax_1}{2}\right)\frac{\gamma t_1^{\gamma-1}}{\Gamma(\gamma+1)}\right]t_1$$
$$+ \left[\frac{a^3}{2}Sinh\left(\frac{\eta}{2} + \frac{ax_1}{2}\right)Sech^3\left(\frac{\eta}{2} + \frac{ax_1}{2}\right)\frac{\gamma(\gamma-1)t_1^{\gamma-2}}{\Gamma(\gamma+1)}\right]\frac{t_1^2}{2!}$$
$$+ \left[\frac{a^3}{2}Sinh\left(\frac{\eta}{2} + \frac{ax_1}{2}\right)Sech^3\left(\frac{\eta}{2} + \frac{ax_1}{2}\right)\frac{\gamma(\gamma-1)(\gamma-2)t_1^{\gamma-3}}{\Gamma(\gamma+1)}\right]\frac{t_1^3}{3!} + \cdots$$

$$v(x_1, t_1) = -1 + \frac{1}{2}a^2 Sech^2\left(\frac{\eta}{2} + \frac{ax_1}{2}\right)$$
$$+ \frac{a^3}{2}Sinh\left(\frac{\eta}{2} + \frac{ax_1}{2}\right)Sech^3\left(\frac{\eta}{2} + \frac{ax_1}{2}\right)\left(\frac{\gamma t_1^\gamma}{\Gamma(\gamma+1)}\right)\left[1 + \sum_{n=1}^{\infty}\frac{(\gamma-n)!}{(n+1)!}\right]$$

$$v(x_1, t_1) = -1 + \frac{1}{2}a^2 Sech^2\left(\frac{\eta}{2} + \frac{ax_1}{2}\right) + \frac{a^3}{2}Sinh\left(\frac{\eta}{2} + \frac{ax_1}{2}\right)Sech^3\left(\frac{\eta}{2} + \frac{ax_1}{2}\right)\left(\frac{\gamma t_1^\gamma}{\Gamma(\gamma+1)}\right) \qquad (2.8.8)$$

Now fixed values $a = \eta = 0.5, \alpha = 1$ and fixed order $\gamma = 1$ in (2.6.7), (2.6.8)
Table: *Solution of LNAM for different values of g when $h = 0.001$ and Absolute Error (AE) of Example 2*

x_1	x_1	$\gamma = 0.55$		$\gamma = 1$		$Ex(\gamma = 1)$		AE	
		u_{LNAM}	v_{LNAM}	u_{LNAM}	v_{LNAM}	u_{Ex}	v_{Ex}	$u_{Ex} - u_{LNAM}$	$v_{Ex} - v_{LNAM}$
-10	0.1	0.0173252	0.008738	0.017556	0.008778	0.017661	0.002791	1.05×10^{-4}	5.8×10^{-3}
	0.3	0.0170452	0.008661	0.017343	0.008671	0.017659	0.002787	3.2×10^{-4}	5.82×10^{-3}
	0.5	0.0168449	0.008606	0.017130	0.008565	0.017657	0.002784	5.27×10^{-4}	5.78×10^{-3}
0	0.1	0.198412	0.098801	0.197184	0.098589	0.196617	0.031096	5.67×10^{-4}	6.74×10^{-2}
	0.3	0.199907	0.099212	0.198316	0.099157	0.196628	0.031114	1.68×10^{-3}	6.80×10^{-2}
	0.5	0.200975	0.099505	0.199451	0.099725	0.196640	0.031131	2.81×10^{-3}	6.85×10^{-2}
10	0.1	0.002515	0.001246	0.002485	0.001240	0.002466	0.000390	1.9×10^{-5}	8.5×10^{-4}
	0.3	0.002555	0.001257	0.002512	0.001257	0.002466	0.001256	4.6×10^{-5}	1×10^{-4}
	0.5	0.002584	0.122656	0.002543	0.001271	0.002467	0.000391	7.6×10^{-5}	8.8×10^{-4}

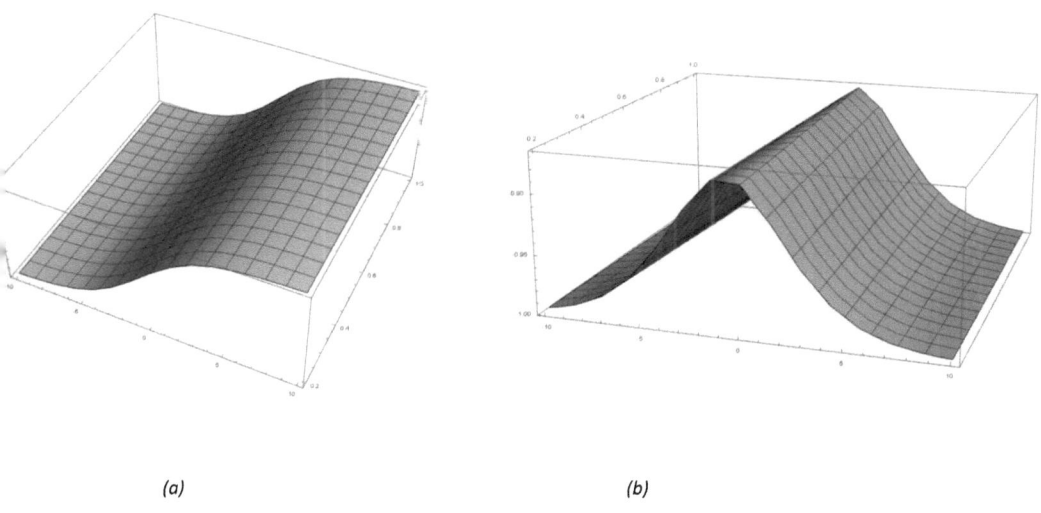

(a)   (b)

Solution of (a) $u(x_1, t_1)$ and (b) $v(x_1, t_1)$ of Example 2 at $\gamma = 1$

Example 3. *Consider the nonlinear KdV of time − fractional order.*

$$\frac{\partial^\gamma u}{\partial t_1^\gamma} = 6u\frac{\partial u}{\partial x_1} - \frac{\partial^3 u}{\partial x_1^3} \ , \qquad\qquad 0 < \gamma < 1 \qquad\qquad (2.8.9)$$

With initial conditions

$$u(x_1, 0) = -2Sech^2(x_1).$$

Taking Laplace transform of Equations (2.6.9)

$$\mathcal{L}\left[\frac{\partial^\gamma u}{\partial t_1^\gamma}\right] = \mathcal{L}\left[6u\frac{\partial u}{\partial x_1} - \frac{\partial^3 u}{\partial x_1^3}\right]$$

$$s^\gamma \mathcal{L}[u(x_1, t_1)] - s^{\gamma-1}[u(x_1, 0)] = \mathcal{L}\left[6u\frac{\partial u}{\partial x_1} - \frac{\partial^3 u}{\partial x_1^3}\right]$$

Applying inverse Laplace transform

$$u(x_1, t_1) = \mathcal{L}^{-1}\left[\frac{u(x_1, 0)}{s} + \frac{1}{s^\gamma}\mathcal{L}\left[6u\frac{\partial u}{\partial x_1} - \frac{\partial^3 u}{\partial x_1^3}\right]\right]$$

$$u(x_1, t_1) = -2Sech^2(x_1) + \mathcal{L}^{-1}\left[\frac{1}{s^\gamma}\mathcal{L}\left[6u\frac{\partial u}{\partial x_1} - \frac{\partial^3 u}{\partial x_1^3}\right]\right]$$

Applying new novel analytic method

For a_1 of $u(x_1, t_1)$

$$a_1 = \frac{\partial}{\partial t_1}(u(x_1, t_1)) = \frac{\partial}{\partial t_1}\left[u(x_1, 0) + \mathcal{L}^{-1}[\frac{1}{s^\gamma}\mathcal{L}\left[6u\frac{\partial u_0}{\partial x_1} - \frac{\partial^3 u_0}{\partial x_1^3}\right]]\right]$$

$$a_1 = \frac{\partial}{\partial t_1}[u(x_1, 0)] + \frac{\partial}{\partial t_1}\left[+\mathcal{L}^{-1}[\frac{1}{s^\gamma}\mathcal{L}\left[6u\frac{\partial u_0}{\partial x_1} - \frac{\partial^3 u_0}{\partial x_1^3}\right]]\right]$$

$$a_1 = \frac{\partial}{\partial t_1}[-2Sech^2(x_1)] + \frac{\partial}{\partial t_1}\left[-16Sech^2(x_1)\frac{t_1^\gamma}{\Gamma(\gamma+1)}\right]$$

$$a_1 = 0 + \left[-16Sech^2(x_1)\frac{\gamma t_1^{\gamma-1}}{\Gamma(\gamma+1)}\right]$$

22

$$a_1 = -16Sech^2(x_1)\frac{\gamma t_1^{\gamma-1}}{\Gamma(\gamma+1)}$$

Similarly For a_2 of $u(x_1, t_1)$

$$a_2 = -16Sech^2(x_1)\frac{\gamma(\gamma-1)t_1^{\gamma-2}}{\Gamma(\gamma+1)}$$

Similarly For a_3 of $u(x_1, t_1)$

$$a_3 = -16Sech^2(x_1)\frac{\gamma(\gamma-1)(\gamma-2)t_1^{\gamma-3}}{\Gamma(\gamma+1)}$$

$\vdots \qquad\qquad \vdots$

$\vdots \qquad\qquad \vdots$

$\vdots \qquad\qquad \vdots$

$\vdots \qquad\qquad \vdots$

Putting the value of $a_1, a_2, a_3 \dots \dots \dots \dots$ in taylor series

$$u(x_1, t_1) = -2Sech^2(x_1) + \left[-16Sech^2(x_1)\frac{\gamma t_1^{\gamma-1}}{\Gamma(\gamma+1)}\right]t_1 + \left[-16Sech^2(x_1)\frac{\gamma(\gamma-1)t_1^{\gamma-2}}{\Gamma(\gamma+1)}\right]\frac{t_1^2}{2!}$$
$$+ \left[-16Sech^2(x_1)\frac{\gamma(\gamma-1)(\gamma-2)t_1^{\gamma-3}}{\Gamma(\gamma+1)}\right]\frac{t_1^3}{3!} + \dots$$

$$u(x_1, t_1) = -2Sech^2(x_1)\left[1 + \frac{8\gamma t_1^\gamma}{\Gamma(\gamma+1)}\left(1 + \sum_{n=1}^{\infty}\frac{(\gamma-n)!}{(n+1)!}\right)\right]$$

$$u(x_1, t_1) = -2Sech^2(x_1)\left[1 + \frac{8\gamma t_1^\gamma}{\Gamma(\gamma+1)}\right]$$

Solution of $u(x_1, t_1)$ of Example 3 ஷ $\gamma = 1$

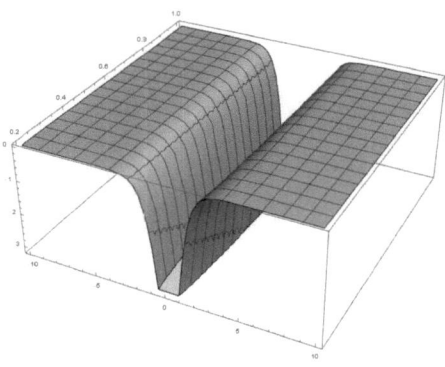

CONCLUSIONS

The objective of our study was to introduce the concept of Fractional Calculus; the branch of Mathematics which explores fractional integrals and derivatives. We first gave some basic techniques and functions, such as the Gamma function and Laplace transformation.

We applied the Laplace–New novel analytical Method for the solution of the fractional KdV type system of partial differential equations. The fractional derivatives are represented by the Caputo operator. The outcome of the proposed method is obtained for both fractional and integer order problems successfully.

Moreover, the behavior of the method is explained through graphs of different numerical examples. The analysis has confirmed that the results obtained by this method are in good contact with the exact solutions for the problems.

The suggested technique is a well-organized methodology with good accuracy and convergence and a strong tool to find approximate analytic solutions for the nonlinear system problems. The tests confirm the validity of a novel technique to handle current nonlinear FPDEs.

In the later, this research can be extended to the investigation by applying this technique for more complicated problems such as systems of nonlinear PDEs.

REFERENCES

1. Shah, R.; Khan, H. Kumam, P. Arif, M: An Analytical Technique to Solve the System of Nonlinear Fractional Partial Differential Equations. Mathematics 2019, 7, 505

2. Ahmed K. Al-Jaberi, Ehsan M. Hameed, Mohammed S. Abdul-Wahab: A novel analytic method for solving linear and nonlinear Telegraph Equation. CoRR abs/2010.02633 (2020)

3. Juan J. Morales-Ruiz, Sonia L. Rueda, Maria-Angeles Zurro: Factorization of KdV Schrödinger operators using differential subresultants. Adv. Appl. Math. 120: 102065 (2020).

4. Lei Li, Jian-Guo Liu: A Generalized Definition of Caputo Derivatives and Its Application to Fractional ODEs. SIAM J. Math. Anal. 50(3): 2867-2900 (2018)

5. Jerry L. Bona, Min Chen, Jean-Claude Saut: Boussinesq Equations and Other Systems for Small-Amplitude Long Waves in Nonlinear Dispersive Media. I: Derivation and Linear Theory. J. Nonlinear Sci. 12(4): 283-318 (2002)

6. V. M. Ashwin, Kumar Saurabh, Sriram Krishnan Murali Krishnan, P. M. Bagade, M. K.

7. Parvathi, Tapan K. Sengupta: KdV Equation and Computations of Solitons: Nonlinear Error Dynamics. J. Sci. Comput. 62(3): 693-717 (2015)

8. Gang-Wei Wang, Xi-qiang Liu, Ying-yuan Zhang: Lie symmetry analysis to the time fractional generalized fifth-order KdV equation. Commun. Nonlinear Sci. Numer. Simul. 18(9): 2321-2326 (2013)

9. David Chiron, Frederic Rousset: The KdV/KP-I Limit of the Nonlinear Schrödinger Equation. SIAM J. Math. Anal. 42(1): 64-96 (2010)

10. George A. Anastassiou: Generalized Fractional Calculus - New Advancements and Applications. Springer 2021, ISBN 978-3-030-56961-7, pp. 1-498

11. Andrea Giusti: General fractional calculus and Prabhakar's theory. Commun. Nonlinear Sci. Numer. Simul. 93: 105114 (2020)

INDEX